IT'S TIME TO LEARN ABOUT IT'S TIME TO LEARN ABOUT HOW TO PASS COLLEGE CHEMISTRY

It's Time to Learn about It's Time to Learn about How to Pass College Chemistry

Walter the Educator

Silent King Books
A WhichHead Entertainment Imprint

Copyright © 2024 by Walter the Educator

All rights reserved. No part of this book may be reproduced in any manner whatsoever without written permission except in the case of brief quotations embodied in critical articles and reviews.

First Printing, 2024

Disclaimer

The author and publisher offer this information without warranties expressed or implied. No matter the grounds, neither the author nor the publisher will be accountable for any losses, injuries, or other damages caused by the reader's use of this book. Your use of this book acknowledges an understanding and acceptance of this disclaimer.

It's Time to Learn about It's Time to Learn about How to Pass College Chemistry is a collectible little learning book by Walter the Educator that belongs to the Little Learning Books Series. Collect them all and more books at WaltertheEducator.com

IT'S TIME TO LEARN ABOUT HOW TO PASS COLLEGE CHEMISTRY

INTRO

College chemistry is a subject that can challenge even the most dedicated students, but with the right approach and mindset, success is achievable. It is not just about memorizing the periodic table or balancing equations, chemistry requires critical thinking, problem-solving skills, and a deep understanding of abstract concepts. For students who are stepping into this field for the first time, it can seem overwhelming. However, by applying strategic learning techniques, mastering foundational principles, and adopting effective study habits, you can navigate college chemistry successfully. This little book will provide you with an in-depth guide on how to not only survive but also thrive in your college chemistry courses.

It's Time to Learn about How to Pass College Chemistry

Understanding the Importance of Chemistry in College

Chemistry is often regarded as the "central science" because it connects and underpins concepts in biology, physics, environmental science, and even engineering. It plays a significant role in various academic disciplines and is often a required course for students in STEM (science, technology, engineering, and mathematics) programs. Understanding the principles of chemistry is essential for anyone pursuing careers in medicine, pharmacology, engineering, environmental science, and more. Even if chemistry is not part of your career goals, the analytical and problem-solving skills you develop through the study of chemistry will serve you in many other areas of life.

It's Time to Learn about How to Pass College Chemistry

Given its central role in education and its complexity, it is crucial to develop strategies that will allow you to succeed. Below, we will explore key strategies to help you pass college chemistry.

It's Time to Learn about How to Pass College Chemistry

Build a Strong Foundation with Chemistry Basics

The first step to succeeding in chemistry is to ensure that you have a solid grasp of the basic concepts. Chemistry is cumulative, meaning each topic builds on the previous ones. Without understanding foundational ideas, it becomes difficult to keep up with more advanced material. Below are some essential areas you need to master:

It's Time to Learn about How to Pass College Chemistry

1. **Atomic Structure:** Atoms are the basic units of matter, and understanding their structure—protons, neutrons, electrons, and how they are arranged—is fundamental. This knowledge will help you grasp chemical bonding, reactivity, and molecular structures.

2. **Periodic Table:** Familiarity with the periodic table is essential. It is not just a collection of elements but a map that organizes elements by their atomic properties. Learn how to use it to predict trends such as electronegativity, ionization energy, and atomic radius, which can explain chemical behaviors.

3. **Chemical Bonding:** Understanding ionic, covalent, and metallic bonds is crucial. You will need to know how atoms combine to form molecules and compounds, as well as how these bonds affect the properties of substances.

4. **Stoichiometry:** This involves calculations related to chemical reactions, including determining the amounts of reactants and products. Mastering stoichiometry requires strong math skills and a clear understanding of the laws of conservation of mass.

5. **Balancing Chemical Equations:** Being able to balance chemical equations accurately is a skill you'll use in almost every topic in chemistry. This ensures that you understand how atoms are rearranged in reactions without violating the conservation of matter.

6. **Acids, Bases, and pH:** The concepts of acidity, basicity, and the pH scale are fundamental to understanding chemical reactivity, especially in organic chemistry and biochemistry.

Before delving into more advanced topics, review these areas thoroughly. If you encounter difficulty in any of these areas, seek help early on, either from professors, tutors, or online resources.

Develop Effective Study Habits

Studying chemistry effectively involves more than just reading a textbook. You need to actively engage with the material in a variety of ways to reinforce learning and promote deeper understanding. Below are some study habits that will enhance your chemistry learning process:

1. **Consistent Practice:** Chemistry is not a subject you can cram for the night before the exam. Regular practice is key. Set aside dedicated study time each day to review notes, solve practice problems, and reinforce concepts.

2. **Active Note-Taking:** Simply copying down what the professor says in class won't suffice. Active note-taking involves summarizing concepts in your own words, drawing diagrams, and highlighting key points. This practice will help cement your understanding.

3. **Use Multiple Resources:** Don't rely solely on your textbook. Explore online resources like Khan Academy, YouTube channels, and other educational platforms. Different explanations of the same topic can provide clarity and deeper insights.

4. **Create Flashcards:** Chemistry involves a lot of terminology, equations, and concepts that you'll need to memorize. Flashcards are a great way to test your knowledge and improve recall. You can use apps like Quizlet for digital flashcards or make your own physical cards.

5. **Join a Study Group:** Collaborative learning can be very effective in chemistry. By discussing concepts with peers, you can gain new perspectives, solve problems collectively, and clarify any doubts you may have. Just make sure the study group remains focused and productive.

6. **Teach What You Learn:** One of the best ways to reinforce your understanding of a topic is to teach it to someone else. Whether it's a classmate or just explaining it out loud to yourself, this technique can expose gaps in your knowledge and solidify your grasp of the material.

7. **Take Practice Tests:** Taking mock exams under timed conditions helps you get used to the format and pressure of real exams. It also helps identify areas where you need further practice.

Master Problem-Solving Techniques

Problem-solving is at the heart of chemistry. From balancing equations to performing complex calculations, you'll need strong problem-solving skills to succeed. Here are some techniques to improve your problem-solving abilities:

1. **Understand the Problem:** Before diving into calculations or writing out solutions, make sure you fully understand the problem. What is being asked? What information is provided? Are there any keywords or concepts you need to apply? Taking the time to break down the problem can save you from making simple mistakes.

2. **Use Dimensional Analysis:** Dimensional analysis, also known as the factor-label method, is a powerful tool for solving problems in chemistry. It involves converting units using conversion factors, ensuring that your final answer has the correct units. This technique is especially useful in stoichiometry and concentration problems.

3. **Draw Diagrams or Write Out Steps:** For more complex problems, especially those involving reactions or mechanisms, it can help to draw out the situation or write down each step in detail. Visual aids like Lewis structures, reaction pathways, or graphs can make abstract problems more tangible.

4. **Check Your Work:** Always double-check your answers. In chemistry, it's easy to make small errors, such as a misplaced decimal point, which can drastically change your results. Going back through your calculations or re-reading the problem ensures accuracy.

5. **Use Practice Problems:** Chemistry textbooks and online resources are full of practice problems. The more you practice, the more familiar you'll become with the types of problems you're likely to encounter on exams.

Utilize Laboratory Time Wisely

Chemistry is unique in that it involves both theoretical work and hands-on laboratory experiments. Your time in the lab is essential for reinforcing the concepts you learn in class. Here's how to make the most of your lab sessions:

1. **Prepare Ahead of Time:** Before going to the lab, thoroughly read the lab manual and understand the experiment you'll be conducting. Familiarize yourself with the equipment and procedures. This will save time and help prevent mistakes.

2. **Stay Organized:** During the experiment, stay organized by keeping detailed notes of your observations, measurements, and calculations. If you make an error, clearly mark it and correct it so that your data remains accurate.

3. **Ask Questions:** If you're unsure about any part of the experiment, don't hesitate to ask your instructor or lab assistant. Understanding why you're performing each step of the experiment is just as important as carrying it out correctly.

4. **Relate Lab Work to Lecture Material:** The lab is where you can see the principles of chemistry in action. Try to connect your lab work to the concepts you've learned in lectures and textbooks. For example, if you're conducting a titration, think about how it relates to acid-base chemistry and pH.

5. **Review Lab Results:** After completing the experiment, review your data and results to see if they align with your expectations. If something went wrong, try to figure out why. Understanding your mistakes is a valuable part of the learning process.

Time Management and Exam Preparation

Proper time management is crucial for success in college chemistry. Between lectures, lab sessions, assignments, and exams, you need to balance your workload effectively. Here are some time management strategies:

1. **Create a Study Schedule:** Develop a weekly study plan that allocates specific times for reviewing notes, solving problems, and preparing for exams. Stick to this schedule as closely as possible.

2. **Don't Procrastinate:** Chemistry can quickly become overwhelming if you fall behind. Review material regularly rather than cramming before an exam. If you don't understand a concept, address it immediately.

3. **Prioritize Assignments and Tests:** Stay on top of your assignments and prepare for exams in advance. If a test is coming up, focus more on reviewing and practicing the topics that will be covered.

4. **Use Office Hours:** Take advantage of your professor's office hours if you need extra help. They can provide clarification on difficult topics, guide you through challenging problems, and give you tips on how to improve your performance.

5. **Sleep and Rest:** Don't neglect your health. Adequate sleep and rest are important for cognitive function and memory. Overworking yourself can lead to burnout, which will hinder your ability to learn and perform well.

Tackling Organic Chemistry

For many students, organic chemistry is one of the most difficult aspects of college chemistry. Organic chemistry involves the study of carbon-based compounds and their reactions. It requires a deep understanding of molecular structures, reaction mechanisms, and synthesis. Here are some tips for tackling organic chemistry:

1. **Learn the Language:** Organic chemistry introduces a lot of new terminology, including functional groups, stereochemistry, and reaction types. Familiarize yourself with these terms and their meanings early on.
2. **Visualize Structures:** Organic chemistry is highly visual. Learn how to draw and interpret Lewis structures, bond-line diagrams, and 3D molecular models. Visualization is key to understanding how molecules interact and react.
3. **Focus on Mechanisms:** Instead of memorizing individual reactions, focus on understanding the underlying mechanisms. Many reactions follow similar patterns, so if you understand the mechanism, you can apply it to different contexts.
4. **Use Molecular Models:** Physical molecular model kits can help you visualize and manipulate the three-dimensional structures of molecules. This can be especially useful when studying stereochemistry and reaction pathways.
5. **Practice, Practice, Practice:** Organic chemistry is notorious for its complexity, so consistent practice is essential. Work through as many problems as you can, and don't be afraid to seek help if you're struggling.

OUTRO

Passing college chemistry is a challenge, but with the right mindset, effective study habits, and a solid grasp of the foundational concepts, you can achieve success. Remember that chemistry requires both theoretical understanding and practical application. Take advantage of all the resources available to you, textbooks, online tutorials, study groups, and lab time, and always seek help when needed. By staying organized, practicing regularly, and managing your time effectively, you'll be well on your way to mastering chemistry and achieving your academic goals.

ABOUT THE CREATOR

Walter the Educator is one of the pseudonyms for Walter Anderson. Formally educated in Chemistry, Business, and Education, he is an educator, an author, a diverse entrepreneur, and he is the son of a disabled war veteran. "Walter the Educator" shares his time between educating and creating. He holds interests and owns several creative projects that entertain, enlighten, enhance, and educate, hoping to inspire and motivate you. Follow, find new works, and stay up to date with Walter the Educator™

at WaltertheEducator.com

www.ingramcontent.com/pod-product-compliance
Lightning Source LLC
LaVergne TN
LVHW010412070526
838199LV00064B/5269